Humanity and the Universe

Copyright Page

The book is copyrighted for 2019

Humanity and the Universe

By Martin K. Ettington

All Rights Reserved USA 2019

ISBN: 9781096787846

Printed in the United States of America

Humanity and the Universe

Humanity and the Universe

The Universe is a very big place and has many mysteries. Most people don't realize just how mysterious it is. Scientists are stumped by many of the conundrums which Astronomy reveals.

In this book I want to give readers a sense of the grandeur, mystery, and puzzles which the Universe poses to scientists. When you realize how incredible our Universe really is you may be surprised.

We used to think the Sun was the center of the Universe. Now we know our solar system is just one of billions. Most of the matter in the Universe is also missing.

And there must be alien life somewhere—you will see the logic and discoveries about this in later chapters.

I also have interspersed some chapters which cover my beliefs in the overall consciousness of the Universe which should provide some interesting perspectives.

There are some amazing facts about our Universe which will also give you some food for thought.

Overall, I hope this book amazes you about what we know about the Universe today and the incredible mysteries which we are just beginning to explore.

Humanity and the Universe

Humanity and the Universe

Other books by Martin K. Ettington

Spiritual and Metaphysics Books:
Prophecy: A History and How to Guide
God Like Powers and Abilities
Enlightenment for Newbies
Removing Illusions to Find True Happiness
Using the Scientific Method to Study the Paranormal
A Compendium of Metaphysics and How to Guides (Six books together in one volume)
Love From the Heart
The Enlightenment Experience
Learn Your Soul's Purpose
Pursuing Enlightenment
A Modern Man's Search for Truth

Longevity & Immortality:
Physical Immortality: A History and How to Guide
The Commentaries of Living Immortals
Records of Extremely Long Lived Persons
Enlightenment and Immortality
Longevity Improvements from Science
The 10 Principles of Personal Longevity
Personal Freedom & Longevity
Telomeres & Longevity
The Diets and Lifestyles of the Worlds Oldest Peoples
The Longevity Six Books Bundle

Science Fiction:
Out of This Universe
Personal Freedom-Parts 1 & 2
The Psychic Soldier Series:
　Book 1-Himalayan Journey
　Book 2-A Soldier is Born
　Book 3-Fighting For Right
　Book 4-Earth Protector
　Book 5-War on the Astral Plane
The Immortality Sci Fi Bundle

The God Like Powers Series:
Human Invisibility
Invulnerability and Shielding
Teleportation
Psychokinesis
Our Energy Body, Auras, and Thoughtforms
The God Like Powers Series—Volume 1 Compilation

The Yoga Discovery Series:
Yoga-An Ancient Art Form
Hatha Yoga-Helping you Live Better
Raja Yoga-Through the Ages
The Yoga Discovery Package

Business Books:
Creating, Publishing, & Marketing Practitioner Ebooks
Building a Successful Longevity Coaching Business
Why Become a Coach?
The Professional Coaching Success Trilogy

Science and Technology
Aliens and Secret Technology
Designing and Building Space Colonies
Future Predictions By and Engineer & Seer
The Unusual Science & Technology Bundle
The Real Atlantis-In the Eye of the Sahara

Humanity and the Universe

The Longevity Training Series

(A transcription of the online Multimedia Longevity Coaching Training Program)

The Personal Longevity Training Series-Book1-Long Lived Persons
The Personal Longevity Training Series-Book2-Your Soul's Purpose
The Personal Longevity Training Series-Book3-Enable Your Life Urge
The Personal Longevity Training Series-Book4-Your Spiritual Connection
The Personal Longevity Training Series-Book5-Having Love in Your Heart
The Personal Longevity Training Series-Book6-Energy Body Health
The Personal Longevity Training Series-Book7-The Science of Longevity
The Personal Longevity Training Series-Book8-Physical Body Health
The Personal Longevity Training Series-Book9-Avoiding Accidents
The Personal Longevity Training Series-Book10-Implementing These Principles

The Personal Longevity Training Series-Books One Thru Ten

These books are all available in digital and printed formats from my website and on Amazon, Barnes & Noble, and Apple ITunes
Website: http://mkettingtonbooks.com

Humanity and the Universe

Table of Contents

Introduction 1
The Moons of our Solar System 3
The Size of the Universe 5
What is the Age of the Universe? 9
Black Holes & Gravity 15
Having 100 Trillion Years to Grow 19
Theories of Universal Structure 21
The Multiverse 23
Dark Matter and Dark Energy 27
Consciousness of the Universe 29
Panspermia and Life 31
Settling the Earth/Moon System 33
   A Lunar Gateway 33
   Lunar Bases 34
Living in Space 37
   Living in Redwood Forest for Five Years 37
   Bora Bora Two 46
The Drake Equation 49
Are we duplicated somewhere? 51
Planets at other Stars 53
Laser Propulsion 61
Alcubierre Drive 63
Aliens on Earth Today 65
The Future of Humanity 71
Longevity and Immortality 73
The Limits to Knowledge 75
Summary 79
Bibliography 81

Humanity and the Universe

Humanity and the Universe

Introduction

The Universe is a very big place and has many mysteries. Most people don't realize just how mysterious it is. Scientists are stumped by many of the conundrums which Astronomy reveals.

In this book I want to give readers a sense of the grandeur, mystery, and puzzles which the Universe poses to scientists. When you realize how incredible our Universe really is you may be surprised.

We used to think the Sun was the center of the Universe. Now we know our solar system is just one of billions.

Most of the matter in the Universe is also missing.

And there must be alien life somewhere—you will see the logic and discoveries about this in later chapters.

I also have interspersed some chapters which cover my beliefs in the overall consciousness of the Universe which should provide some interesting perspectives.

There are some amazing facts about our Universe which will also give you some food for thought.

Overall, I hope this book amazes you with regard to what we know about the Universe today and the incredible mysteries which we are just beginning to explore.

Humanity and the Universe

Humanity and the Universe

The Moons of our Solar System

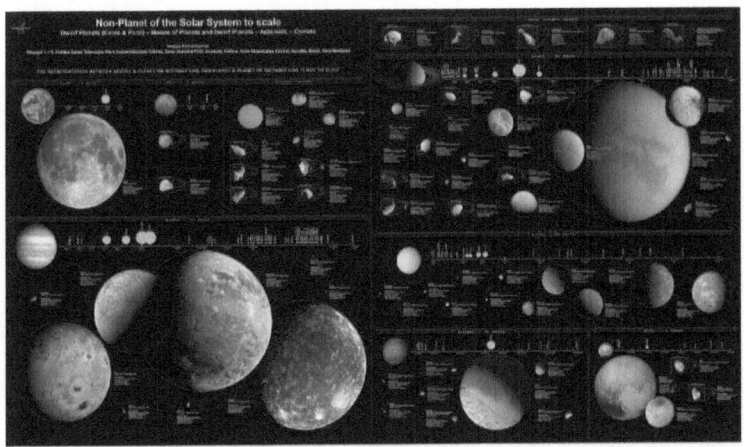

Before we get to the rest of the Universe we should understand that we have a tremendous opportunity to explore and settle our own solar system.

We don't just have eight major planets. (Pluto has been downgraded to a minor planet.) There are over **194 known moons** around these planets. And the number keeps expanding as our telescopes and probes tell us more about our system. Take a look at the table below showing how many moons have been discovered around the planets:

Summary – number of moons

Planet	Mercury	Venus	Earth	Mars	Jupiter	Saturn	Uranus	Neptune
Number of moons	0	0	1	2	79	62	27	14

Humanity and the Universe

In addition to the moons there are hundreds of asteroids in the Asteroid belt, and large objects in the Oort cloud too. (The Oort cloud is at the edge of the Solar System beyond Pluto)

So we have a huge amount of space to settle within our own Solar System—not counting space settlements—which will be discussed later in this book.

Admittedly, these moons will need high technology for settlements and to provide effective living conditions—but we are already planning to do this on the Earth's moon, so the technology will exist soon.

The Size of the Universe

The Universe we know about has grown in incredible size and complexity over the last few centuries.

One of the first cosmological models was the geocentric model developed by the Greek astronomer Ptolemy. Ptolemy's model of the Universe placed the Earth at the center with the sun and planets located in concentric crystal spheres surrounding Earth. These spheres rotated, causing the sun and planets to appear to rise and set. The stars were fixed in a stationary outer sphere. During the

Humanity and the Universe

Middle Ages, this model became widely accepted in Europe, because the central location of Earth reaffirmed the importance of man.

In the 1400s, scientists were beginning to question Ptolemy's model. In his book "On the Revolutions of the Heavenly Spheres", the church canon and astronomer Nicolaus Copernicus proposed a heliocentric model which placed the sun, instead of the earth, at the center of the solar system. Copernicus's model would later be championed by the famed scientist Galileo Galilei.

So the Earth had started to lose its place in men's minds as being the center of the Universe. In the early 20th century this reduction in importance would reach further with the understanding that we live in just one of many galaxies. At the time it was thought that there was just one Galaxy—The Milky Way. The Milky Way was determined to be 100,000 light years across. (A light year is the distance it takes light to travel in one year.) That everything else was just nebulas and clouds of gases outside of ours.

Astronomer Edwin Hubble determined in 1924 that there were Galaxies outside of our own including the now well-known Andromeda Galaxy which is 2.5 million light years away. The number of galaxies known at that time were only a few dozen. These Galaxies then must be the outer limit of the Universe-right?

Over the years, larger telescopes and the Hubble Space telescope have revealed that the number of galaxies in our Universe is immensely larger.

The latest estimates of the total number of galaxies in our observable Universe as of 2018 is **over two trillion galaxies**. This is an order of magnitude larger than the estimated number of stars in the Milky Way Galaxy. The stars in our home galaxy alone are estimated to be over 200 billion. This means that the number of stars in the Universe should be around 4 to the 21st power of stars. This is 4,000,000,000,000,000,000,000 individual stars.

We also know that there might be more galaxies beyond the observable horizon—so far out that we can never see the light from them.

Currently, scientists estimate the diameter of the size of the Universe to be over 91 billion light years. How do we comprehend such a large number? We know that the edge of the observable Universe is 46 billion light years so 91 is just double that-assuming we are in the center of it. (Are we are back to the Ptolemaic model of the Universe at a much larger scale?—thinking we are at the center of it all)

Our Universe is of an unimaginable size. How do we comprehend it? One way to think of the Universe's size is to recognize that there might be galaxies that have expanded in space beyond the limits of our observable Universe.

Humanity and the Universe

What is the Age of the Universe?

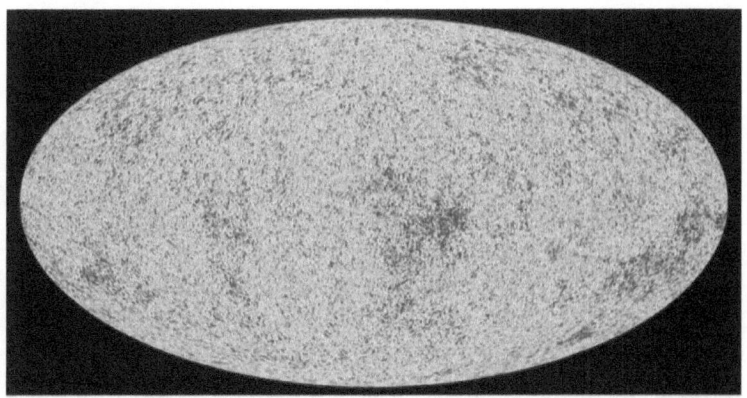

So how old is the Universe? Most scientists today believe in the Theory of "The Big Bang" which is the moment that the Universe grew from a single point in an incredible explosion.

The age of the Universe is measured different ways by different methods but most methods agree that it is around 13.8 Billion years old.

Various methods have been used to come to this determination. Here are some of them:

Observational limits

Since the universe must be at least as old as the oldest things in it, there are a number of observations which put a lower limit on the age of the universe; these include the temperature of the coolest white dwarfs, which gradually cool as they age, and the dimmest turnoff point of main sequence stars in clusters (lower-mass stars spend a greater amount of time on the main sequence, so the lowest-mass stars that have evolved off of the main sequence set a minimum age)

Cosmological parameters

The age of the universe can be determined by measuring the Hubble constant today and extrapolating back in time with the observed value of density parameters (Ω). Before the discovery of dark energy, it was believed that the universe was matter-dominated (Einstein–de Sitter universe, green curve). Note that the de Sitter universe has infinite age, while the closed universe has the least age.

The value of the age correction factor, F, is shown as a function of two cosmological parameters: the current fractional matter density Ωm and cosmological constant density $\Omega \Lambda$. The best-fit values of these parameters are shown by the box in the upper left; the matter-dominated universe is shown by the star in the lower right.

The problem of determining the age of the universe is closely tied to the problem of determining the values of the cosmological parameters. Today this is largely carried out in the context of the ΛCDM model, where the universe is assumed to contain normal (baryonic) matter, cold dark matter, radiation (including both photons and neutrinos), and a cosmological constant. The fractional contribution of each to the current energy density of the universe is given by the density parameters Ωm, Ωr, and $\Omega \Lambda$. The full ΛCDM model is described by a number of other parameters, but for the purpose of computing its age these three, along with the Hubble parameter {\displaystyle H_{0}} H_{0}, are the most important.

Humanity and the Universe

WMAP

NASA's Wilkinson Microwave Anisotropy Probe (WMAP) project's nine-year data release in 2012 estimated the age of the universe to be $(13.772\pm0.059)\times10^9$ years (13.772 billion years, with an uncertainty of plus or minus 59 million years).

However, this age is based on the assumption that the project's underlying model is correct; other methods of estimating the age of the universe could give different ages. Assuming an extra background of relativistic particles, for example, can enlarge the error bars of the WMAP constraint by one order of magnitude.

*****

So we have a pretty good idea about how long our Universe has existed. The next question then is how long will our Universe live? This is pretty important since humanity would have a hard time existing if the Universe was dead.

Theories on the future life of the Universe vary from 5 billion to 100 trillion years or even more. If the truth is near the upper end of this range then Humanity has been born near the beginning of the Universe's creation. How amazing would that me?

Maybe Humanity will become the super advanced civilization that others will look to in the future. That would mean **humanity was born in one hundredth of one percent of the beginning of the Universe**. We will explore this idea in a later chapter.

There are various theories about the Universe's life depending on whether our Universe is expanding, contracting, or in a steady state. These theories can be summarized as follows:

Closed universe

If Omega >1, then the geometry of space is closed like the surface of a sphere. The sum of the angles of a triangle exceeds 180 degrees and there are no parallel lines; all lines eventually meet. The geometry of the universe is, at least on a very large scale, elliptic.

In a closed universe, gravity eventually stops the expansion of the universe, after which it starts to contract until all matter in the universe collapses to a point, a final singularity termed the "Big Crunch", the opposite of the Big Bang. Some new modern theories assume the universe may have a significant amount of dark energy, whose repulsive force may be sufficient to cause the expansion of the universe to continue forever—even if Omega >1

Open universe

If Omega <1, the geometry of space is open, i.e., negatively curved like the surface of a saddle. The angles of a triangle sum to less than 180 degrees, and lines that do not meet are never equidistant; they have a point of least distance and otherwise grow apart. The geometry of such a universe is hyperbolic.

Even without dark energy, a negatively curved universe expands forever, with gravity negligibly slowing the rate of expansion. With dark energy, the expansion not only continues but accelerates. The ultimate fate of an open universe is either universal heat death, the "Big Freeze", or the "Big Rip", where the acceleration caused by dark energy eventually becomes so strong that it completely overwhelms the effects of the gravitational, electromagnetic and strong binding forces.

Conversely, a negative cosmological constant, which would correspond to a negative energy density and positive pressure, would cause even an open universe to re-collapse to a big crunch. This option has been ruled out by observations.

Flat universe

If the average density of the universe exactly equals the critical density so Omega =1}, then the geometry of the universe is flat: as in Euclidean geometry, the sum of the angles of a triangle is 180 degrees and parallel lines continuously maintain the same distance. Measurements from Wilkinson Microwave Anisotropy Probe have confirmed the universe is flat with only a 0.4% margin of error.

In absence of dark energy, a flat universe expands forever but at a continually decelerating rate, with expansion asymptotically approaching zero. With dark energy, the expansion rate of the universe initially slows down, due to the effect of gravity, but eventually increases. The ultimate fate of the universe is the same as an open universe.

*****

So depending on whether the Universe is Open, Closed, or Flat this will also have an effect on how long our Universe may exist.

Humanity and the Universe

Black Holes & Gravity

Back in the early twentieth century Albert Einstein wrote his Theory of Special Relativity in 1905 and the more expanded General Relativity Theory in 1915 which also covered Gravity.

Later theorists found an interesting result of his theories in that they calculated than some stars of certain size could shrink to a point when collapsing at the end of their lives. (Stars larger than the Swartzchild Radius)

This was an amazing proposition and was considered just a nice theory for many decades. In 2015 Astronomers actually found evidence of the existence of Black Holes. Wow—something so amazing which was just theorized actually exists.

> *A black hole is a region of space time exhibiting such strong gravitational effects that nothing—not even particles and electromagnetic radiation such*

as light—can escape from inside it. The theory of general relativity predicts that a sufficiently compact mass can deform space time to form a black hole. The boundary of the region from which no escape is possible is called the event horizon. Although the event horizon has an enormous effect on the fate and circumstances of an object crossing it, no locally detectable features appear to be observed. In many ways a black hole acts like an ideal black body, as it reflects no light. Moreover, quantum field theory in curved space time predicts that event horizons emit Hawking radiation, with the same spectrum as a black body of a temperature inversely proportional to its mass. This temperature is on the order of billionths of a kelvin for black holes of stellar mass, making it essentially impossible to observe.

The implications of black holes are huge. One thought is that they are gateways to another universe, or to the substrate lying underneath our Universe where there is no time and space.

Personally, I like the concept of the underlying substrate to our Universe being outside of space and time. This would mesh nicely with the concepts of the spirit and our souls which are supposed to live in a timeless and space less realm.

Another implication is the idea of travelling through black holes to another location in space and avoiding relativistic speed requirements. (If you can avoid dying when entering a black hole's event horizon.)

Humanity and the Universe

Here is a new picture of a black hole. It is thought there might be black holes in the center of every galaxy and each consists of millions of stars:

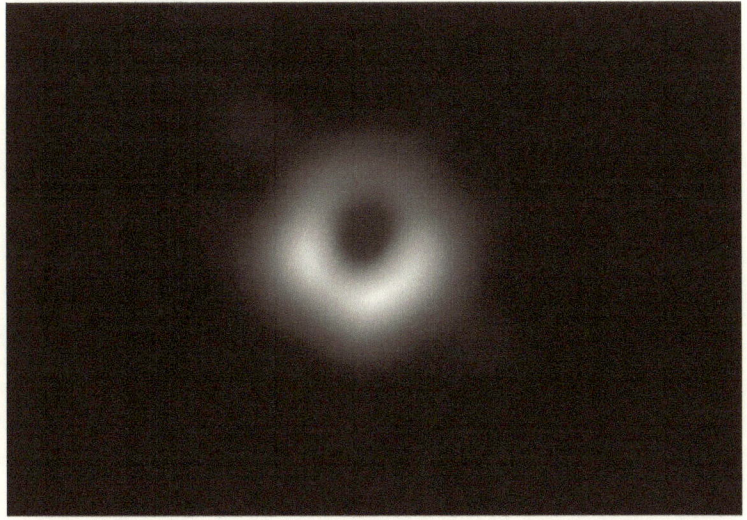

Humanity and the Universe

Having 100 Trillion Years to Grow

Consider if the Universe could last another 100 trillion years—Although there might be many other civilizations millions of years in advance of Humanity, as a race, we would still be one which has started near the beginning of our Universe's life.

What do I mean by the beginning Of the Universe? Because scientists estimate that the current universe is 13.7 billion years old. That seems pretty old.

Our own Solar System was formed over 4.6 billion years old. This is an unimaginable age. So why do I say our race exists near the beginning of the Universe? This is since some theories on the end of the Universe put its potential length at **one hundred trillion years**. (A recent theory says the end will be in five billion years—but most estimates are much longer)

Humanity and the Universe

One Hundred Trillion years. That is an incredible amount of time. By that estimate the current Universe is only at 0.00014 of that age or only .014 percent. Wow—if that's true humanity is relatively one of the first intelligent species in the Universe.

What will we become in 100 trillion years? I know this is an unimaginable amount of time. And who knows if the human race or our Super God like descendants will survive that long.

Biological evolution would probably only go on for another billion years or so, so we might be talking about evolution of the mind and spirit, or maybe mechanical and computer evolution. (Assuming we don't just modify our own genes directly and forget about Natural Selection)

Maybe most of us would live in artificial computer generated realities in a billion years or…

Our decedents might become a race of Gods compared to us and a guiding force of the Universe.

I know that dealing with concepts of this amount of time are really impossible to conceive. We would have a hard enough time predicting how beings only a few hundred years more advanced than us would act. So what incredible things could happen to our race over thousands, millions, or even billions of years?

Theories of Universal Structure

The Universe as we currently know it is broken down into voids and filaments. These are composed of Super Clusters of Galaxies at the high end, then Clusters, Galaxy Groups, and finally the individual Galaxies.

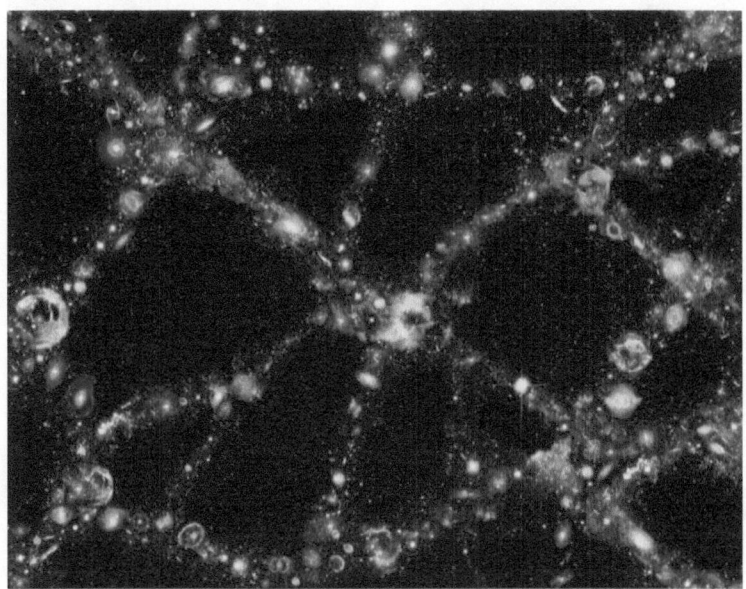

Above is an illustration of Super Clusters of Galaxy groups near the Milky Way. A Super Cluster might be composed of 100,000 individual Galaxies stretched across hundreds of millions of light years.

But we don't need to go up to that level to be amazed by the complexity of numbers of spatial components around us.

Humanity and the Universe

Let's just look at the spiral arms of our galaxy and we will be awed by how big everything really is:

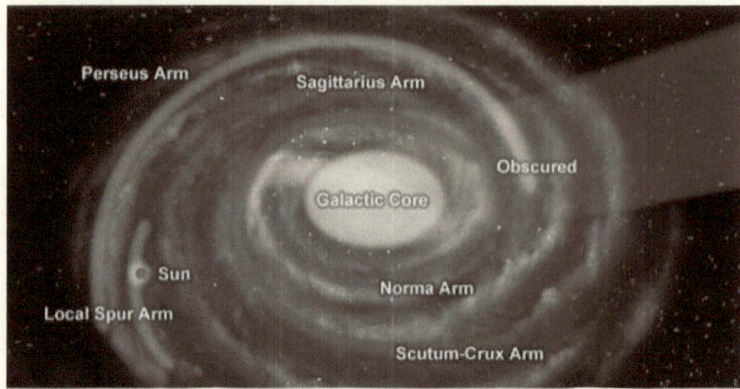

The above picture shows the known arms of our galaxy and remember that there are estimated to be over 200 billion stars just in the Milky Way Galaxy.

Humanity and the Universe

The Multiverse

There are many theorists who have propounded the theory that there are other Universes which may exist simultaneously with our Universe. These theories are called the "Multiverse" theories.

Here is an extract from an article which describes these ideas from a more technical perspective:

> The structure of the multiverse, the nature of each universe within it, and the relationships among these universes vary from one multiverse hypothesis to another.
>
> Multiple universes have been hypothesized in cosmology, physics, astronomy, religion, philosophy, transpersonal psychology, and literature, particularly in science fiction and fantasy. In these contexts, parallel universes are also called

"alternate universes", "quantum universes", "interpenetrating dimensions", "parallel dimensions", "parallel worlds", "parallel realities", "quantum realities", "alternate realities", "alternate timelines", "alternate dimensions", and "dimensional planes".

The physics community continues to debate the multiverse hypotheses. Prominent physicists are divided in opinion about whether any other universes exist.

Some physicists say the multiverse is not a legitimate topic of scientific inquiry. Concerns have been raised about whether attempts to exempt the multiverse from experimental verification could erode public confidence in science and ultimately damage the study of fundamental physics. Some have argued that the multiverse is a philosophical rather than a scientific hypothesis because it cannot be falsified. The ability to disprove a theory by means of scientific experiment has always been part of the accepted scientific method. Paul Steinhardt has famously argued that no experiment can rule out a theory if the theory provides for all possible outcomes.

In 2007, Nobel laureate Steven Weinberg suggested that if the multiverse existed, "the hope of finding a rational explanation for the precise values of quark masses and other constants of the standard model that we observe in our Big Bang is doomed, for their values would be an accident of the particular part of the multiverse in which we live.

Think about this—what if every decision we make or action which occurs could also happen in a separate Universe but

Humanity and the Universe

with alternative decisions. An example might be when you drive to a fork in the road. In our Universe you might make a left turn while that might become a right turn in another Universe. Maybe the left turn ends the trip safely and the right turn results in a car accident later on. This might mean the history of each Universe branches out into multiple possibilities as events occur. This might mean an infinite set of Universes.

Or what about an entirely difference approach to this question. What if there were other Universes created with different scientific constants than ours. Maybe Gravity is a stronger—or weaker force than ours. What if the speed of light is different? Life might not be possible in these other Universes, or maybe those conditions are more beneficial to life. There are an infinite number of possibilities.

Life might exist only in a subset of these Universes because the physical laws in some may not support life at all. Other Universes might be even more supportable of life than ours.

Humanity and the Universe

Dark Matter and Dark Energy

(Blue colors indicate volumes which might hold dark matter)

Astronomers had a problem—the galaxies were rotating faster than they should have for the number of stars observed. Different observers had different theories:

> *The hypothesis of dark matter has an elaborate history. In a talk given in 1884, Lord Kelvin estimated the number of dark bodies in the Milky Way from the observed velocity dispersion of the stars orbiting around the center of the galaxy. By using these measurements, he estimated the mass of the galaxy, which he determined is different from the mass of visible stars. Lord Kelvin thus concluded that "many of our stars, perhaps a great majority of them, may be dark bodies". In 1906 Henri Poincaré in "The Milky Way and Theory of*

> *Gases" used "dark matter", or "matière obscure" in French, in discussing Kelvin's work.*

The mystery remained unexplained and more detailed observations were made in the 1960s and 1970s:

> *Vera Rubin and Kent Ford in the 1960s and 1970s provided further strong evidence, also using galaxy rotation curves. Rubin worked with a new spectrograph to measure the velocity curve of edge-on spiral galaxies with greater accuracy. This result was confirmed in 1978. An influential paper presented Rubin's results in 1980. Rubin found that most galaxies must contain about six times as much dark as visible mass; thus, by around 1980 the apparent need for dark matter was widely recognized as a major unsolved problem in astronomy.*

Many expensive experiments looking for dark matter have been conducted over the last twenty years and none of them have shown any results.

So some theorists are proposing that there are other weak fields around the low density parts of galaxies which could account for the faster rotations of them than the known matter would provide.

Dark Matter and Dark Energy are still a great mystery about the construction of our Universe.

Consciousness of the Universe

Is the Universe conscious? A strange question you might say...

Most westerners think of the Universe as totally a physical construct. Matter and Energy, and maybe hidden Dark Matter and Dark Energy. But consciousness?

Well, according some belief systems, every speck of matter has some portion of the infinite consciousness of the Universe.

According to Theosophical beliefs not only is our Mother Earth or "Gaia" conscious, but so is the "Solar Logos" at the next level.

Beyond that level is the consciousness of the Galaxy, and then larger constructions of galaxies, until you get to the overall consciousness of the Universe or what we call GOD.

Humanity and the Universe

Many Hindus believe that every component of matter is conscious. That each rock has consciousness. It's an interesting concept. One can sit and contemplate a stone, and try to feel the way the stone feels about everything going on around it. I've tried it and it leads to some interesting thoughts.

You can also sit at the base of a tree and contemplate the tree's consciousness. I've done this sitting below large ancient trees—like Redwoods or Sequoias and you get a real sense of the majesty of what that consciousness of an ancient being would feel.

As stated in an earlier chapter about Black Holes, I do think there is a strong argument for our spirits or souls being part of a timeless and space less realm.

One reason I believe our spirit exists in this realm is because of my premonitional experiences. These events could not occur if we didn't have some aspect of our consciousness which exists outside of time and can see the future.

See my book "Prophecy: A History and How to Guide" for more details on my prophecy experiences.

Panspermia and Life

Panspermia is the term used to describe the concept of Earth's life being populated naturally by life from other stars.

We know that spores can float on rocks and meteors in space for thousands if not millions of years. It's certainly feasible that some of those rocks might have landed on the Earth over those millennia and spores inside of them might have lived through reentry into the atmosphere.

Now there is more evidence of life having existed inside of meteorites. A research team in Hungary has found another tantalizing clue – mineralized and filament-like organic material embedded in a Martian meteorite – ALH-77005 – from the Allan Hills region of Antarctica. The material in the meteorite is similar to that produced by iron-oxidizing microbes on Earth. See the below picture showing the mineralized location of the organic material:

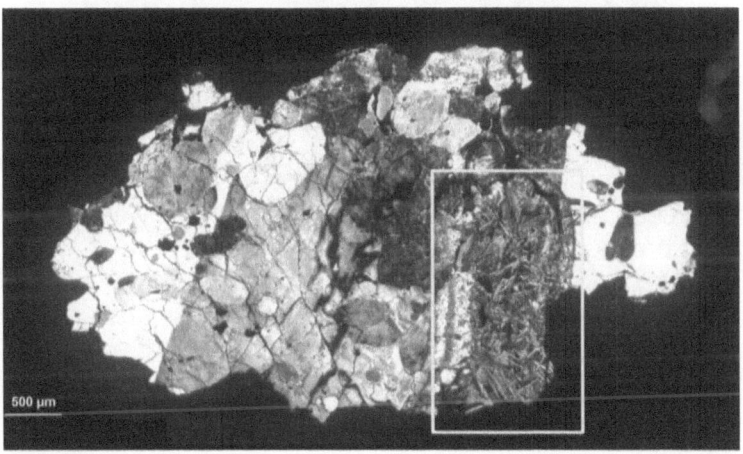

Humanity and the Universe

In the last couple of decades, scientists have studied and come to realize that life on Earth can also live below ground to the depth of miles—and inside what we normally consider to be inanimate rocks.

As of late 2018 scientists now estimate that there are 16.5 to 25 billion tons of life living below the earth. This is a tremendously huge number and might really explain how come there is so much oil underground. (Since Geologists believe that oil is the result of generations of the decomposition of living organisms.)

Below is the picture of a nematode found deep in the Kopanang Gold Mine in South Africa nearly one mile below the surface.

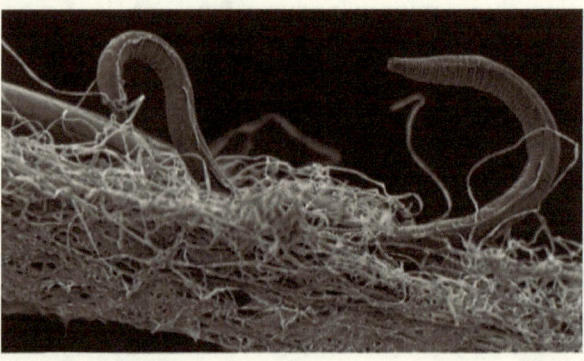

If life can exist at such depths in our own world, what does that mean about the viability of life travelling between the stars embedded in large rocks or asteroids?

This might mean that life can exist inside rocks and asteroids for many thousands or even millions of years.

Humanity and the Universe

Settling the Earth/Moon System

Man likes to explore. We enjoy challenging the unknown, and with the immense size of our Universe, we can barely wait to get started. Here are some of the next likely steps in Humanity moving out into space:

<u>A Lunar Gateway</u>

Humanity's knowledge of technology for space travel and living in space is growing rapidly. NASA is already planning a small space station to orbit the moon and looking at what will be needed to travel back to the moon and live there.

With ice being known to exist at the Moon's poles, this could allow oxygen and fuel to be mined and refined to help support the existence of a manned moon base.

Here is a proposal for what a moon orbiting base or "Gateway" might look like:

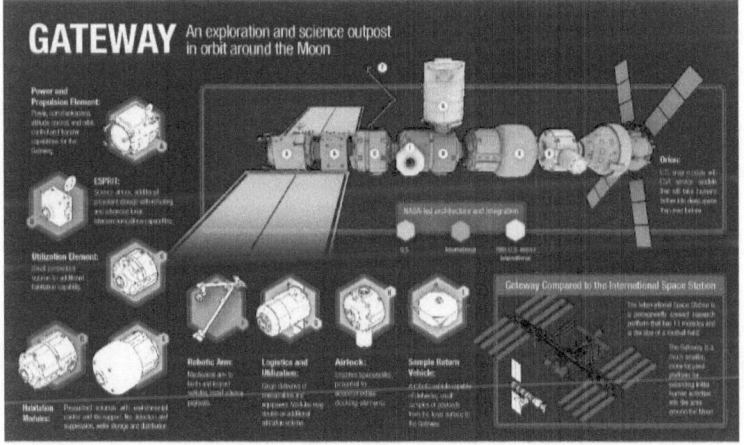

Humanity and the Universe

This gateway would have multiple modules and have an oblong orbit which goes close to the moon and far away. This type of orbit lets spacecraft going to and from the gateway to have easier orbits to connect with it and the Moon.

There would be a habitation module, an airlock, refueling module, and logistics module, among others.

A lunar lander could be fueled at the Lunar Gateway to reach the Moon's South Pole. Later it would return to the gateway and be re-fueled for another trip.

Lunar Bases

NASA has also made requests for proposals of moon landing ships, different experiments, moon bases, and much more.

A moon base at Shakelton Crater on the South Pole of the moon would allow explorers to accomplish several things:

Humanity and the Universe

Try to use the ice which has been identified there to produce fuel for spaceships and for human usage.

To build a small base which can be manned to give us knowledge of how to build and maintain a base remote from Earth like we would need to inhabit Mars.

The base in Shakelton Crater would likely be built from an inflatable dome, which is then covered by Lunar Regolith.

The covering of regolith would insulate the dome from radiation, micro-meteoroids, and reinforce the pressure structure to keep an atmosphere and heat inside.

Humanity and the Universe

Living in Space

I've written some Science Fiction books about building space settlements. My main novel on this subject is titled "Personal Freedom-Parts 1 & 2" Part 1 is about building a large O'Neil type space settlement. The settlement consists of two connected counter rotating cylinders. Here are some chapters which give a sense of how living in this space settlement might be:

Living in Redwood Forest for Five Years

After five years of building our Habitat community this is how things were:

Humanity and the Universe

The manufacturing companies had shrunk and sold off most of their land areas so they now only occupied one eighth of the Redwood Forest Habitat.
They were also building their own manufacturing facilities outside our Habitats and within the L5 point to move all of the dangerous processes.

All of the practices which were a danger to the Habitat or residents would soon be removed to safer areas.
The rest of the Forest Habitat had undergone full initial landscaping. You could stand on the "outside" in the Habitat and see many lakes and hills, with grassed covering them, and trees in the process of building forests. The tallest trees were now thirty feet high and you could get an idea of what the forests would eventually look like.
Multiple cities had grown and they each consisted of thousands of residents. You could see towers in many of them—since there was plenty of room to grow—which could grow even up to two and a half miles high.

Imagine standing at the End Cap in zero gravity where you enter the Redwood Forest Habitat on the inside.
Near the entrance you would see stairs going down into the ground that led the Habitat transportation system.
There were also a couple of trails which led out of the entrance and down into deeper gravity areas.

The trails started as concrete curving path, in and around trees and grass as you descended. After a mile you started feeling the increase gravity and there were some railings on the trail to help you stabilize and get used to the increased gravity. (Which had been set to one gravity at the highest from increased rotational speeds several years ago.)

There were even some areas to lie on the grass and look upward.

Humanity and the Universe

When you did that you saw the entire inside of the Habitat and it was glorious. Imagine three strips of land with lakes interspersed and towns and cities going all the way to the other Hub—until they faded away in the distance. The air being moist and there being haze which turned the other end into blurs.

In between the three strips of living area were the windows where the solar mirrors shined in simulating day and night. There were also gradual gradations of light as the day progressed.

It all had many colors and looked like you were in an airplane looking down on everything.

The Redwoods and other trees were all still less than thirty feet, so the forests were still small, but you could tell what they would become by looking at the overall plan.

We even had a river with rapids where you could test yourself Kayaking if you wanted.

*****

Some new sports had arisen too. Since the gravity near the centerline of each Habitat was zero, it was possible launch yourself from the End Caps and fly along the centerline.

Many people learned how to fly and even held soccer like games in freefall. With floating goalposts. Teams were formed and it became a great televised competition.
In case you got too much out of the centerline and were brought down by gravity, you could open a parachute and float down safely.

Humanity and the Universe

This new living space led Sharon-one of our physical trainers-- to start a new really tough competition.
She was adapting the idea of the Triathlon to our Habitat. Sharon called it the "Greek Space Marathon". (I don't know why—it just sound good)

The first race was amazing and many of the dual Habitat settlers took breaks from work to stand at the sidelines or watch on monitors.

I watched too since I was really excited and curious about how the competitors would navigate the challenges.
The distances were also scaled down since Sharon added a few segments.

The field of competitors at the starting line were thirty people—men and women, since some of the segments needed skills; not just endurance and speed.
The first segment started in the main habitat area and was a twenty mile bike ride in one gravity.

The men pulled ahead of most of the women, but a few women were faster than the lagging men.

The second segment was using Kayaks to go down the river. A few tipped over and those people were fished out and were out of the race too.

Women caught up at this point until the remaining twenty five competitors were evenly interspersed-men and women.

I could see that Sharon with her yellow shirt was racing and she was in the middle.
The third segment was a three mile swim in one of the lakes. A couple competitors were already exhausted and had to be pulled from the water since they could not go on.

Humanity and the Universe

The fourth event, with a now reduced field, were twenty one persons and it was a race to the end cap from the lake.

The race was fifteen miles but it was a very unusual course. The race started as normal, but the gravity kept going down. 80% at mile three, 50% at mile seven, and gravity kept reducing from there.

In the last five miles the competitors weren't running as much as jumping. In one tenth of a gravity some jumps could go up twenty five feet.

This is where the real skill came in and a lot of women started catching up. If you jumped too high, it would take too long to land and others would pass you. You had to jump on the right trajectory. Not just use force.

More people came down at bad angles and got sprained ankles or generally banged up and had to quit.
This left fifteen competitors at the end cap for the fifth challenge.

I saw that the group included Sharon, but also some younger teens—boys and girls, who had been raised in the Habitat the last five years and were the closest we had to "Natives". These kids had lots of time to practice and they were used to variable gravities.
The fifth challenge was to fly down the centerline of the Habitat for five miles using a pair of wings you would flap through a mechanical linkage to your arms and legs.
Once you reached the end of the line—which was a banner held into place by tiny jets, you would parachute to the ground.

Humanity and the Universe

The sixth and final challenge was the skydive, and the first one to land in the target area on the ground was the winner.

Women started to catch up on the men, again because skills were needed to flap the wings and their lighter mass made a difference in flying down the centerline.
What we didn't expect in this first race was that the teenagers started catching up and passing the adults with much more strength and endurance.

As the group got to the end of their flying challenge, we saw them dropping off the wings and skydiving down to the finish.

Sharon was still in the race, although the real competitors had dropped to ten—since some didn't know how to fly well and were lagging far behind.

So it was Sharon and the teenagers-five boys and girls who dropped fast without opening their chutes. They were hoping to gain speed and open their chutes at the end to win.

I should also mention that each competitor wore a safety harness with steam powered safety thrusters. If they went too fast and the computer monitoring sensors decided they were going to crash, it would slow them way down to a safe landing, but they would also be out of the race. Several of the kids were way too enthusiastic and their thrusters went off—keeping them safe, but they were fouled out of the race.

It was a close finish, but when it was over a girl named Marra who was fifteen years old had won the race, and Sharon came in second.

Humanity and the Universe

Only seven people finished the complete race. The rest had been thrown out in various race segments or injured on the way.

There were several hundred people at the party celebrating this inaugural race, and Marra got the Winners Crown—smiling shyly as many teen girls do.
Even those who got injured were sitting down—with their bandages on-- and having a great time with the fresh barbeque and wine.

It seemed that a new sport had been invented and everyone started making plans to attend or compete in next years "Greek Space Race".

*****

Amazingly, one our biggest cash producing industries was tourism. Visitors would stay at new hotels—we even had some hotel chains from Earth, and they would spend a few weeks hiking around, boating in the lakes, skiing, sailing, or generally relaxing.

But one of the most exciting things the tourists did which really excited me, was to attend classes on the 10 Principles. They wanted to learn how to live to their optimum potential and personal freedom, and take what they learned back with them to change their Earth habits. I could see that teaching people the 10 Principles and how to use them in their lives could have a very long term impact on the rest of humanity.

These Tourists also brought in a lot of money too which didn't hurt our community at all.

*****

Humanity and the Universe

Our people continued to practice mindfulness techniques and integration of the 10 Principles into their lives.

The full impact of our communities learning the 10 Principles in their lives were just becoming known to us. Once thing was a lot less religious strife. People brought their own religions with them and they had been building churches, synagogues, and mosques.

But, since these people realized the commonalities of all religions, they were polite together and many interfaith events were planned. Everyone had respect for others cultures and religions.

We had our disagreements-as expected, but everyone in our communities were involved in how to reach a consensus on issues of importance.

Schools were working out, and children were learning important skills in them and having fun too.

In fact one of the newest initiatives was to start planning a college—one for each habitat.

One was to be an engineering and technical school concentrating on construction technologies and techniques we had learned in building the Habitats.
A key difference in the Engineering School was what and how we teach.

We taught construction not only with basic courses, but with a virtual construction program where students learned to build habitats of their designs while wearing virtual reality suites.

The programs were constructed using videos and computer simulations we took from the actual construction.

Humanity and the Universe

Students could use tools as appropriate and even speed up the clock in simulation to see the results from their efforts.

Many a student died virtually while building in space, but they learned a lot in the process.

The other things students learned was how to apply the 10 Principles to their lives as an engineer and working in the construction environment.

They learned how being more spiritually centered would help them to remain sharp and objective in a dangerous construction environment, and they learned about how to develop consensus during meetings by avoiding emotions and looking to the common good and the best decisions objectively.

The other college in Bora Bora Two was to be about Philosophy and Wellness; combining the practical application of the 10 Principles to people's lives.

That school was designed to produce coaches and teachers who had internalized these principles and wanted to help others.
Side courses taught specifics for implementation including Yoga of different types, Meditation Instruction, Tai Chi Teaching, and more.

Humanity and the Universe

Bora Bora Two

Talking about Bora Bora Two is fun because I love that place!

Jennifer and I would often take weekend "vacations" there with our friends to see it

We took the tube trains which now connected distant parts of the Redwood Forest Habitat with each other as well as to Bora Bora Two.

These electric and computer controlled trains could take you to stations anywhere in the Habitats within 30 minutes. Trains came to each station every 15 minutes and could each hold 50 people.

We reached Bora Bora Two (BB2) on the train from our station and headed out into the "islands" to the beach house we usually stayed at.

Inside BB2 it was 70% water, with a simulated ocean near the shore. The Ocean was continuing to be stocked with salt water fish and other sea animals including many types of fish and shellfish.

A reef was even in the process of being built and grown to provide a real reef experience to our children and more diversity of life to our Habitat.

Streams were pumped out at the top of hills to have fresh water and waterfalls everywhere too.

The Ocean was up to 100 feet deep for bottom dwelling fish to live and to have a darker experience for some breeds of fish and whales.

Humanity and the Universe

Yes-we had whales—a small breeding colony of grey whales, but they added to our diversity and the food chain. We also had canals to let the whales move from segment to segment to let them move through the seas and all over BB2.

If you hiked some of the islands—which went up to over 1000 feet high-with cliffs, and looked out over the Ocean, you got a real feel for being on a South Seas Island in the Pacific.

We did avoid raising man eating sharks, but did have some of the gentler breeds as part of the overall food chain.

*****

Several fish farms and fishing businesses developed to supply residents and visitors with fresh seafood for their diets.

You could go to a fish market or a restaurant and find many of the choices an earthly fish market would offer. So imagine us living on the beach outside of our house there, and grilling lobsters along with fresh swordfish steaks together.

Add to that some fresh corn on the cob which we brought with us from Redwood Forest, and some home grown beer, and you have the makings of a little summer feast. We sat with our friends after being satiated by dinner and watch the manmade twilight descend over the ocean. The number of colors on the ocean were amazing.

It was incredible to think that we had created this new world from nothing and in the dark and cold emptiness of space.

Humanity and the Universe

People were meant to live in tune with Nature. Even if it was an artificially created one.
But our sense of adventure was not totally fulfilled. What could we do next for a next adventure?

Humanity and the Universe

The Drake Equation

In this chapter we explore what science and astronomy can tell us about the likelihood of alien races in our galaxy. (The Milky Way)

a. The Drake equation

Dr. Frank Drake an astronomer proposed an equation on the possibility of intelligent life at the first scientific meeting on the search for extraterrestrial life in 1960. (SETI). This equation has since become the definitive calculation for the existence of alien life.

The Drake equation is shown below with definitions of its variables.

$$N = R^* \, f_p \, n_e \, f_l \, f_i \, f_c \, L$$

- N = The number of communicative civilizations
- R^* = The rate of formation of suitable stars (stars such as our Sun)
- f_p = The fraction of those stars with planets. (Current evidence indicates that planetary systems may be common for stars like the Sun.)
- n_e = The number of Earth-like worlds per planetary system
- f_l = The fraction of those Earth-like planets where life actually develops
- f_i = The fraction of life sites where intelligence develops
- f_c = The fraction of communicative planets (those on which electromagnetic communications technology develops)
- L = The "lifetime" of communicating civilizations

Once you fill in the variables this equation will tell give you an estimate of the number of civilizations on other planets that we should be able to communicate.

Now that we are finding lots of planets and expect to find many planets that can support life, the probability of intelligent life has gone up significantly.

When filling in the variables based on conservative assumptions and our current knowledge of others stars, we find that there should be thousands of intelligent civilizations in just our own galaxy to talk to.

b. What Astronomy and Astrophysics tells us

A good place to start is with what our current understanding of science and astronomical observations of the Universe tell us is possible.

Most Astronomers calculate that the age of our Universe is between 13.5 to 14 billion years old. That there are also galaxies and stars that were formed only a billion years after the big bang that started our universe.

Life on earth took several billion years to start and our planet is estimated to be 4.5 billion years old. Therefore, if life exists on any planets in the universe it may have started as much as nine to ten billion years old.
In the 1990s our telescopes and astronomy tools also became good enough to start finding planets around other stars. As of 2017, 4,500 candidate exoplanets have been found around stars by the Kepler Observatory. Of these about ten are Earth like planets in that they are in the stars habitable zone and are rocky planets like Earth. Almost every star examined seems to have planets.

We also know that there are at least 100 billion stars in our galaxy, and maybe more. (There are an estimated two trillion galaxies overall.) Let's also make an assumption that there might be one out of one thousand Earth like worlds which have intelligent life. Given the above information collected from Kepler and other observatories we can conclude the following:

That there should be over 100 million planets in our Galaxy alone which support intelligent life. That is a lot of potential alien civilizations.

Humanity and the Universe

Are we duplicated somewhere?

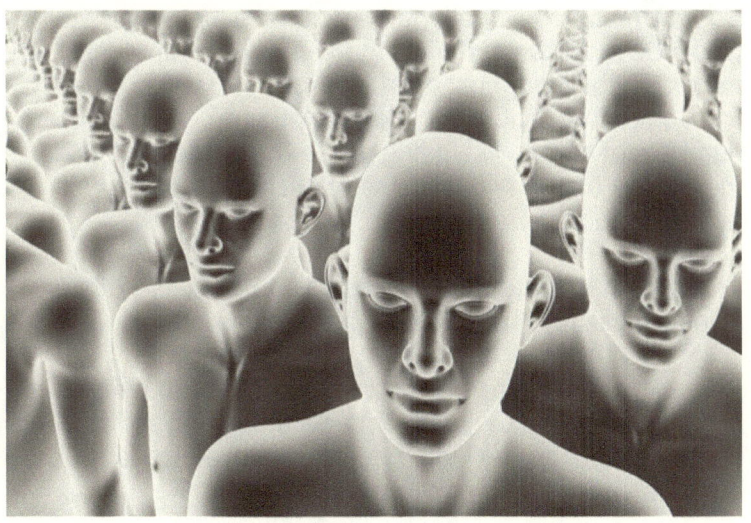

One of the thoughts I've had as I'm writing this book is that there might be many other duplicates of myself throughout this Universe. Here is my logic:

Scientists estimate that there are over 2 trillion galaxies in the Universe.

We can calculate that this means there must be over 4,000,000,000,000,000,000,000 individual stars.

Now let's assume that only one out of a million stars has intelligent life on it. (This is probably low from what we know about planets around stars.) This still means there would be 4,000,000,000,000,000 stars with intelligent life. If only one millionth of those stars has human like life, and a couple of billion people per planet, we are still looking at

Humanity and the Universe

4,000,000,000,000,000,000 individual beings of human like life in the universe.

So how many combinations of genes are there which would be just like each of us in expression of our looks and minds? I would suggest even a billion—but it is not an infinite number. There is no known number, but the overall numbers of human like beings in the Universe says that there must be many duplicates of each of us on planets going around different stars.

These numbers are really impossible to calculate with any scientific validity, but I strongly believe this is true.

Sometimes I just close my eyes and imagine that my spirit is connecting with these other duplicates of myself around the universe. The feeling I get from this exercise is immensely satisfying and relaxing.

My thought is that these genetic duplicates of myself somewhere else in the Universe should be on the same wavelength as myself and therefore we are able to sense each other.

Humanity and the Universe

Planets at other Stars

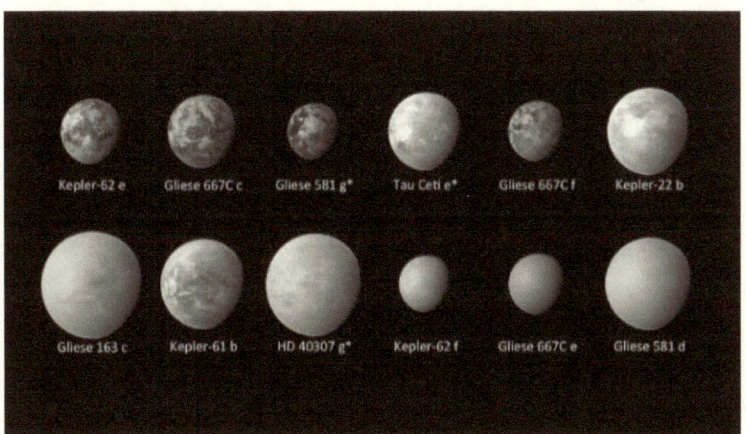

The latest findings from our space probes are that almost every star has planets around it. This is a big change from when I was growing up in the middle of the twentieth century. At that time Astronomers thought that planets might be rare in the Galaxy and only found around certain types of stars.

Think about this—there must be an incredible number of planets out there. And the estimates from current observations of stars by the Kepler spacecraft and other observatories says that at least one in ten should have an earth like planet in the habitable zone around the star.

Wow! There are potentially billions of habitable plants in our galaxy. An uncountable number of places where life might exist. And at least millions of planets which might host intelligent life.

The Kepler spacecraft has now finished its observations of other stars and found over 4,500 planets around various

Humanity and the Universe

stars. But most of those stars are not close to Earth. The new spacecraft planet hunter is called TESS and has the following goal:

> *TESS stars will be 30-100 times brighter than those surveyed by the Kepler satellite; thus, TESS planets should be far easier to characterize with follow-up observations. These follow-up observations will provide refined measurements of the planet masses, sizes, densities, and atmospheric properties.*

The planets found by TESS should be in our local star neighborhood. After some new planets are found they will be observable to determine more detail by the new James Webb Space Telescope soon to be launched.

Humanity and the Universe

Humanity Travelling to Other Stars

Eventually—and I mean hundreds of years from now; and probably well after the Solar System is well settled, we may want to visit other stars and even plant colonies on the planets around them.

So the question is how to get there. One of the earliest science fiction approaches to answering this question was to build a generational ship. There are many stories where the persons living on the generational ship for centuries, have descendants who have not been educated to the fact that they are even on a ship.

By building an O'Neil colony as discussed previously, and adding rockets to it, there may very well be the possibility to travel to other stars over hundreds or even thousands of years.

But would anyone signup for such a long journey? It's likely that persons already living in a space colony might signup since they were planning to live out their lives there anyway.

Generational Ships

The following issues would need to be considered for a generational ship to travel successfully to the nearest star:

Biosphere

Such a ship would have to be entirely self-sustaining, providing energy, food, air, and water for everyone on board. It must also have extraordinarily reliable systems that could be maintained by the ship's inhabitants over long periods of time. This would require testing whether thousands of humans could survive on their own before sending them beyond the reach of help. Small artificial closed ecosystems, such as the Biosphere 2, have been built in an attempt to work out the engineering difficulties in such a system, with mixed results.

<u>Biology and society</u>

Generation ships would have to anticipate possible biological, social and morale problems, and would also need to deal with matters of self-worth and purpose for the various crews involved. As an example, a moral quandary exists regarding how intermediate generations, those destined to be born and die in transit without actually seeing tangible results of their efforts, might feel about their forced existence on such a ship.

Estimates of the minimum reasonable population for a generation ship vary. Anthropologist John Moore has estimated that, even in the absence of cryonics or sperm banks, a population capacity of 160 people would allow normal family life (with the average individual having ten potential marriage partners) throughout a 200-year space journey, with little loss of genetic diversity; social engineering can reduce this estimate to 80 people. In 2013 anthropologist Cameron Smith reviewed existing literature and created a new computer model to estimate a minimum reasonable population in the tens of thousands. Smith's numbers were much larger than previous estimates such as Moore's, in part because Smith takes the risk of accidents and disease into consideration, and assumes at least one severe population catastrophe over the course of a 150-year journey.

In light of the multiple generations that it could take to reach even our nearest neighboring star systems such as Proxima Centauri, further issues on the viability of such interstellar arks include:

- The possibility of humans dramatically evolving in directions unacceptable to the sponsors

- The minimum population required to maintain in isolation a culture acceptable to the sponsors; this could include such aspects as

- The ability to maintain and operate the ship

- The ability to accomplish the purpose (planetary colonization, research, building new interstellar arks) contemplated

Sharing the values of the sponsors, which may not be likely to be empirically demonstrated to be viable beyond the home planet unless, once the ship is away from Earth and on its way, survival of one's offspring until the ship reaches the target star is one motivation.

Size

In order for a spacecraft to maintain a stable environment for multiple generations, it would have to be large enough to support a community of humans and a fully recycling ecosystem. However, a spacecraft of such a size would require a lot of energy to accelerate and decelerate. A smaller spacecraft, while able to accelerate more easily and thus make higher cruise velocities more practical, would reduce exposure to cosmic radiation and the time for malfunctions to develop in the craft, but would have challenges with resource metabolic flow and ecologic balance.

Social breakdown

Generation ships travelling for long periods of time may see breakdowns in social structures. Changes in society (for example, mutiny) could occur over such periods and may prevent the ship from reaching its destination. This state was described by Algis Budrys in a 1966 book review:

The slower-than-light interstellar spaceship, pursuing its way through the weary centuries, its crew losing touch with all reality save the interior of the vessel ... Well, you know the story, and its unhappy downhill round, it's exciting struggles between the barbarian tribes which develop in its disparate compartments, and then, if the writer is so minded, the ultimate flash of hope as the good guys win out and prepare to meet their future on some noble, if erroneous basis.

Robert A. Heinlein's Orphans of the Sky (the "impeccable statement of this theme", Budrys said) and Brian Aldiss's Non-Stop (U.S. title Starship) discussed such societies.

Cosmic rays

Health threat from cosmic rays. The radiation environment of deep space is very different from that on the Earth's surface, or in low earth orbit, due to the much larger influx of high-energy galactic cosmic rays (GCRs). Like other ionizing radiation, high-energy cosmic rays can damage DNA and increase the risk of cancer, cataracts, and neurological disorders. One known practical solution to this problem is surrounding the crewed parts of the ship with a thick enough shielding such as a thick layer of maintained ice as proposed in The Songs of Distant Earth, a science fiction novel by Arthur C. Clarke (note: in this book the ship's mammoth ice shield is only in the forward part of the ship, preventing micrometeors from damaging the ship during its interstellar journey).

Technological progress

If a generation ship is sent to a star system 20 light years away, and is expected to reach its destination in 200 years, a better ship may be later developed that can reach it in 50 years. Thus, the first generation ship may find a century-old human colony after its arrival at its destination.

Humanity and the Universe

Laser Propulsion

Lasers are a feasible method of sending a probe or spaceship to nearby stars. The idea is to shine a high power laser at the construction's solar sail to propel it at high speed to that star. Here is more of a description in detail:

A key piece of technology for interstellar travel may be within our reach, according to US scientists, but a plan to send a laser-propelled spacecraft to the star Proxima Centauri is still a long way from reality.

Thanks to the mind-boggling size of space, humans have so far been limited to exploring our own backyard. Proxima Centauri is the Sun's closest neighbor, but reaching it with current rocket technology would take approximately 12,000 years.

In 2016, the Breakthrough Starshot initiative suggested an alternative: use massive laser arrays on Earth to propel a fleet of miniature, ultra-light space probes. These probes would be attached to lightsails: large sheets of reflective material that are driven by photons rather than wind.

Travelling at 20% of the speed of light — a blistering pace of over 200 million km/hr — the probes would reach Proxima Centauri in just 20 years, soaring through the system and beaming home data, including images of the potentially Earthlike planet Proxima b.

Alcubierre Drive

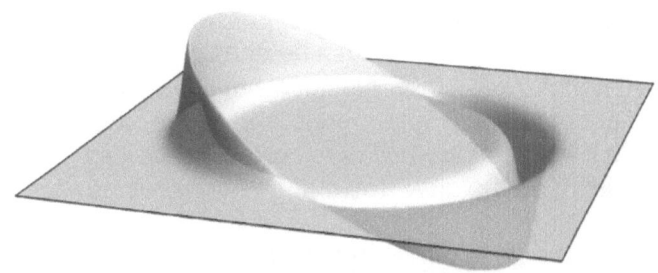

The Alcubierre Drive is a proposed faster than light type of spaceship propulsion which would not violate Einsteinian Physics. Here is more of a description of the concept:

The Alcubierre drive or Alcubierre warp drive (or Alcubierre metric, referring to metric tensor) is a speculative idea based on a solution of Einstein's field equations in general relativity as proposed by Mexican theoretical physicist Miguel Alcubierre, by which a spacecraft could achieve apparent faster-than-light travel if a configurable energy-density field lower than that of vacuum (that is, negative mass) could be created.

Rather than exceeding the speed of light within a local reference frame, a spacecraft would traverse distances by contracting space in front of it and expanding space behind it, resulting in effective faster-than-light travel. Objects cannot accelerate to the speed of light within normal space time; instead, the Alcubierre drive shifts space around an object so that the object would arrive at its destination faster than light would in normal space without breaking any physical laws.

Humanity and the Universe

Although the metric proposed by Alcubierre is consistent with the Einstein field equations, it may not be physically meaningful, in which case a drive will not be possible.

Even if it is physically meaningful, its possibility would not necessarily mean that a drive can be constructed. The proposed mechanism of the Alcubierre drive implies a negative energy density and therefore requires exotic matter. So if exotic matter with the correct properties cannot exist, then the drive could not be constructed.

However, at the close of his original article Alcubierre argued (following an argument developed by physicists analyzing traversable wormholes) that the Casimir vacuum between parallel plates could fulfill the negative-energy requirement for the Alcubierre drive.

Another possible issue is that, although the Alcubierre metric is consistent with Einstein's equations, general relativity does not incorporate quantum mechanics. Some physicists have presented arguments to suggest that a theory of quantum gravity (which would incorporate both theories) would eliminate those solutions in general relativity that allow for backwards time travel (see the chronology protection conjecture) and thus make the Alcubierre drive invalid.

Humanity and the Universe

Aliens on Earth Today

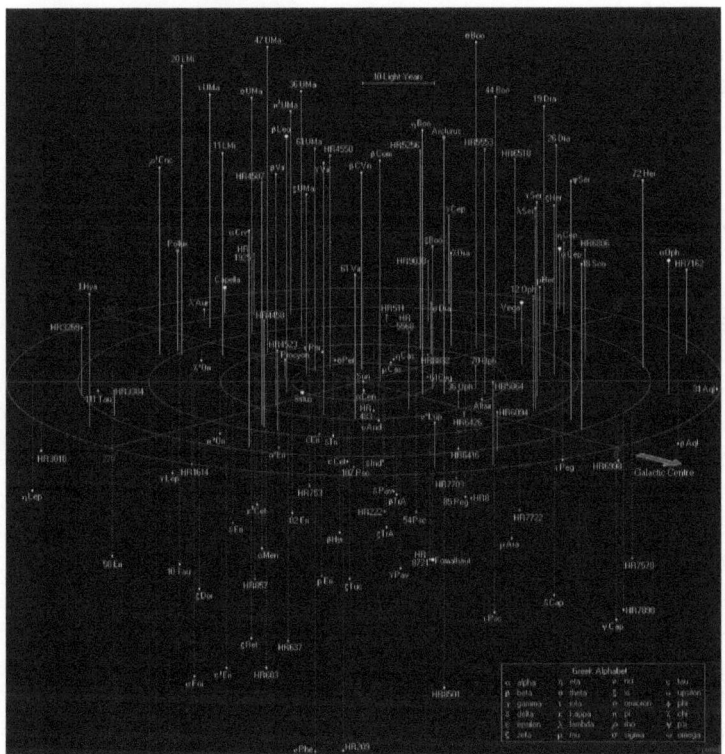

(A picture of stars within 50 light years of Earth is above)

There are over 500 G type stars within 100 light years of Earth. G type stars are similar to our own. There are also over 250,000 stars of all types within 250 light years. Let us assume that we don't need a G type star to produce life. We just need planets in the habitable zone of the star—which is different for different type stars.

Humanity and the Universe

Given that one in ten stars has an earth type planet in the habitable zone then that means there are 25,000 Earth like planets within 250 light years of Earth.

So there might be a good chance for intelligent life within stars close to ours. A lot of candidate planets for aliens are in local space.

Now, what evidence do we have for aliens visiting Earth? More than you might think. A book by a former soldier who worked for NATO said that they had top secret reports there are at least 18 and as many as 100 alien races who had visited Earth. The actual report has not been found and so is controversial.

The U.S. Military and other research organizations have been researching invisibility for several decades now with some success. Any sufficiently advanced civilization would have invisibility technologies which would seem as magic to us. But there might occasionally be glitches and the alien visitors could become visible. Here are some of the best alleged photos of Alien visitors to review…

Below is a photo of kids in a jungle with a possible alien behind them:

Humanity and the Universe

Next is a picture taken in Chile as part of a video which the photographer says is un-doctored and might be an alien who thought he was invisible:

Another picture of an alien sitting on a road:

What about other alien interactions with humanity?
Have you ever heard of Serpo.org? This website claims with lots of detail that aliens from the star Zeta Reticuli contacted the US government and that we actually had interactions with them.

That we sent a group of a dozen astronauts on their spaceship to their planet for over 10 years. There are a lot more details to the story which are quite amazing. This might have been the factual basis for the movie "Close Encounters of the Third Kind".

Again, any alien civilization which is thousands or even just hundreds of years ahead of us would likely have developed invisibility/stealth technology to hide both spaceships and individuals on the Earth. So there might be thousands of aliens currently visiting Earth that we don't know about.

Humanity and the Universe

Humanity and the Universe

The Future of Humanity

So what does the future hold for Humanity?
Will we explore and settle our solar system and eventually the stars?

It's been almost fifty years since man first landed on the moon. I remember the event clearly. I was a teenager at Boy Scout camp and the whole camp went to the dining hall where a TV was setup. We watched Neil Armstrong as he made his first steps on the moon. Everyone was amazed beyond words and we all figured that soon we would have settlements on the Moon and travel to Mars.

But that didn't happen. Instead NASA invested in the Space Shuttle to return to low Earth orbit with no goals beyond building a Space Station-also in low Earth Orbit. The Space Shuttle was a great concept, but because of too many design compromises it was never cost effective. It is estimated that it cost $450 million dollars in 2011 to launch one mission. With the costs of a little more than $100 million for Falcon Heavy launches, we could have launched five missions into Earth Orbit, or several to the moon.

The Space Shuttle was also unfortunately dangerous, with two of them lost in launch and reentry accidents and over twelve crew members killed. Traditional rockets with escape tower rockets and capsule reentry are much safer. Now that rocket technology is becoming re-usable and many more companies are getting into the rocket business, the cost per pound of launching things into orbit should drop precipitously

Humanity and the Universe

Humanity and the Universe

Longevity and Immortality

The Universe is almost infinite in terms of its size and the time it would take to explore everything and all the stars in it is almost infinite—even at speeds faster the light. Our lives are as short as a microbe compared to the time we would need to make a decent attempt at exploration of the stars.

So there is good reason for man to focus on learning how humans can have extreme longevity. I've written a number of books on the topic of Longevity and consider myself an expert.

My belief is that there are no absolute limits to how long a person can live. There are already records of persons living to their mid one hundreds, and some claims of persons living much longer.

The ability of people to live for thousands of years to take advantage of long interstellar travel would make sense when man decides on the almost infinite effort to explore the Universe.

See my books on http://mkettingtonbooks.com for more information. The longevity books include "Physical Immortality: A History and How to Guide" and "The 10 Principles of Personal Longevity"

Humanity and the Universe

Click on the Images below to see the book descriptions:

Humanity and the Universe

The Limits to Knowledge

Our scientific understanding of the world is only about 500 years old. Mankind has existed for over 100,000 years, and the universe is billions of years old.

Is humanity so arrogant as to say that we have a close to final understanding of the natural scientific laws of the universe, or should we be more humble and admit that we only understand a tiny fraction of what is out there, and much more is undiscovered than discovered.

I spent many years reading articles and journals from organizations like the ASPR (American Society for Psychical Research) who have done good experimental work for 60 years on validating and understanding psychic phenomenon. However if I were to go to the average person on the street they would say that these things have never been proven.

(I also read the standard scientific journals like Science and Scientific American.) Most scientists would also say that paranormal events haven't been proven to exist.

Instead of exploring how to understand and benefit from these abilities, most researchers in these areas are still being asked to prove that these things really exist. In this case many of the skeptics aren't really interested in the objective evidence because it would disrupt their cozy worlds.

Humanity and the Universe

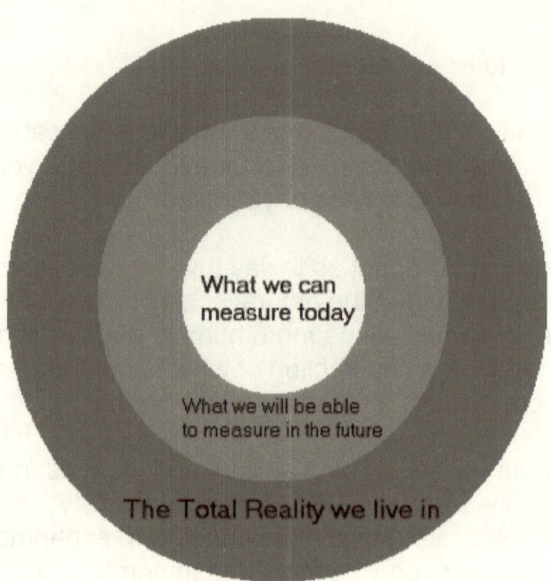

From the foregoing discussion on the scientific method and what is measurable, you can tell that I must have a significantly different idea of reality than the norm. The circles within circles picture above best illustrates my belief of our ability to understand Reality:

The inner yellow circle represents what we can measure with our instruments today and perform experiments on to prove or disprove theories.

The red circle is a larger area, which we will eventually be able to measure to understand and prove or disprove the way things are

The blue outer circle is the largest area, and is that part of the universe which we may be able to experience but will never be able to measure and validate with objective scientific approaches.

Humanity and the Universe

We may be able to subjectively perceive a lot of things in the blue area, but will never have the tools and techniques to objectively quantify it.

This blue realm may also include such things as where the soul goes after death, the fundamental nature of God, and certain dimensions of space and time, which we can postulate but never prove or disprove.

The red region may be more amenable to creative approaches for objective measurement and validation.

However, there will have to be agreement among the scientific community on some new approaches, which may constitute legitimate standards for objective measurement of phenomena.

This may include indirect evidence, which is used in areas like particle physics. Neutrinos for example can't be directly perceived, but their existence can be inferred by collisions with other particles, which make cloud tracks which we can directly perceive.

The same approach should be transferable to validation of something like telepathy, where the medium of thought transference may not be understood at this point, but it can be validated through well-controlled blind studies and statistics.

I think that this type of validation issue of the objectivity of an experiment also presents barriers to further scientific progress.

Humanity and the Universe

Humanity and the Universe

Summary

I hope my exposition on our wonderful Universe has provided some facts which amazed you, and some thoughts to mull over in your free time.

Some of the things which I find most interesting are:

That our known Universe keeps expanding. From just the Solar System hundreds of years ago, to the Galaxy in the early 20th century, then thousands of galaxies, and now 2 trillion galaxies and a possible Multiverse.

When I was growing up the only things we knew about in our Solar System were the planets and their major Moons. Now due to space probes and better telescopes we know of hundreds of individual bodies. There are lots of place to visit in our system.

Although this is a Biology issue and not outer space, it is incredible that life has been found miles below the surface of our planet in rock. This certainly provides a better argument that life could exist and travel across space in rocks.

There are also multiple plausible pictures of Aliens on our Earth. The ideas of Aliens living hidden on our world are also gaining more interest and credibility

Could we have duplicates of ourselves around the Universe? We can't compute the odds because we don't know how often intelligent life occurs or the range of genetic expressions for humanity. However, this is a valid concept.

Humanity and the Universe

Humanity and the Universe

Bibliography

1. https://earthsky.org/space/evidence-fossil-life-martian-meteorite-alh-77005. *earthsky.org.* [Online] 2019.

2. https://en.wikipedia.org/wiki/Age_of_the_universe. *Wikkpedia: Age of the Universe.* [Online]

3. https://en.wikipedia.org/wiki/Ultimate_fate_of_the_universe . *Wikipedia: Ultimate Fate of the Universe.* [Online]

4. https://en.wikipedia.org/wiki/Timeline_of_the_far_future. *Wikipedia: Timeline of the far future.* [Online]

5. http://www.solstation.com/stars3/100-gs.htm. *G Stars within 100 light years of Earth.* [Online]

6. https://www.space.com/24894-exoplanets-habitable-zone-red-dwarfs.html. *Space.com: Nearly every star hosts at least one Alien Planet.* [Online]

7. https://tess.gsfc.nasa.gov/whytess.html. *TESS: Transiting Exoplanet Survey Satellite.* [Online]

8. https://www.bibliotecapleyades.net/exopolitica/esp_exopolitics_zzzzzd.htm. *The Assessment Report on Aliens.* [Online]

9. https://en.wikipedia.org/wiki/Multiverse. *Wikipedia: The Multiverse.* [Online]

10. https://en.wikipedia.org/wiki/Black_hole#Observational_evidence. *Wikipedia: Black Holes.* [Online]

11. https://en.wikipedia.org/wiki/Dark_matter#Early_history. *Wikipedia: Dark Matter.* [Online]

12. https://cosmosmagazine.com/space/laser-powered-travel-to-nearby-stars-a-step-closer. *Space News.* [Online] 2018.

13. serpo.org. *The Serpo Project.* [Online] 2005.

www.ingramcontent.com/pod-product-compliance
Lightning Source LLC
Chambersburg PA
CBHW031923170526
45157CB00008B/3031